KIRIBATI'S CLIMATE RESILIENCE

Kiribati's Climate Resilience

Road to Sustainable Planet

ANURAG ANURAG

Anurag Anurag

Contents

This is a work of fiction. Unless otherwise indicated, all the names, characters, businesses, places, events, and incidents in this book are either the product of the author's imagination or used in a fictitious manner. Any resemblance to actual persons, living or dead, or actual events is purely coincidental.

Chapter 1

Kiribati's Precarious Position

Kiribati is a small island nation located in the central Pacific Ocean. The country consists of 33 atolls and reef islands, which are home to approximately 120,000 people. Unfortunately, Kiribati is facing an existential threat due to climate change-induced rising sea levels.

The average elevation of Kiribati's atolls is only two meters above sea level, making them highly vulnerable to flooding and saltwater intrusion. Experts predict that, by the end of the century, Kiribati's landmass could reduce significantly, putting the lives and homes of its people at risk. This issue is not limited to environmental factors; social, economic, and cultural aspects are also at stake.

The impact of rising sea levels on Kiribati's coastal communities is already visible. Coastal erosion caused by the encroaching tides is causing land loss, endangering communities, infrastructure, and crucial resources like freshwater lenses. The loss of land and homes is a significant concern, as it puts people's lives and livelihoods at risk. In addition, traditional practices associated with the land and ocean are at risk of disappearing, impacting Kiribati's cultural heritage.

For instance, fishing is one of the traditional practices in Kiribati. The rising tides are affecting the availability of fish, which is causing food insecurity for the people of Kiribati. This situation is compounded by the loss of fishing grounds to coastal erosion. Furthermore, navigation is another traditional practice that is under threat. The rising sea levels are causing the islands' shorelines to recede, which is affecting the accuracy of navigation methods that rely on land-based markers.

Despite these challenges, Kiribati's people are demonstrating resilience through community-based adaptation strategies and innovative approaches to cultural preservation. For instance, the government of Kiribati has established a program that involves relocating people from the most vulnerable islands to the less vulnerable ones. This program aims to reduce the risk of loss of life and property due to rising sea levels.

Moreover, the people of Kiribati are taking steps to preserve their cultural heritage. For example, they are developing innovative techniques to preserve traditional knowledge of fishing and navigation. The government is also investing in education to ensure that future generations are aware of their cultural heritage and the challenges they face.

In conclusion, rising sea levels are a significant threat to Kiribati's existence. The country is facing the risk of losing its land, homes, and cultural heritage. However, the people of Kiribati are showing resilience and determination to safeguard their identity and homeland against these challenges. The government is implementing innovative approaches to adaptation and cultural preservation, while the people are developing new techniques to preserve their traditional practices. Kiribati's experience is a reminder of the urgent need to address climate change and its impact on vulnerable communities worldwide.

Chapter 2

Battling Plastic Pollution in Kiribati's Waters

The story of Kiribati is one of resilience, determination, and the pursuit of a sustainable future, even as the island nation faces a range of challenges brought about by climate change. Rising sea levels, extreme weather events, and environmental degradation are all taking a toll on the country, but Kiribati's response has been characterized by innovation, community-driven initiatives, and the blending of traditional wisdom with modern solutions.

One of the most pressing issues facing Kiribati is plastic pollution in the Pacific Ocean surrounding the country. Debris, including single-use plastics and discarded fishing gear, inundates beaches and harms marine life, threatening the ecosystem's balance and impacting Kiribati's traditional fishing practices and the health of its people who rely on seafood. Plastics' long-lasting impact on marine life - from entanglement to ingestion - highlights the urgent need for comprehensive solutions to mitigate the harm and reduce plastics' presence in Kiribati's waters.

Despite the challenges, Kiribati has been proactive in addressing plastic pollution through community-driven initiatives. Beach clean-ups, educational programs, and recycling initiatives have all been developed to tackle the issue. However, limited waste management infrastructure and financial resources present significant challenges, which necessitate collaboration with international organizations and neighboring countries. This collaborative approach is crucial for curbing the inflow of marine debris and implementing sustainable waste management practices.

Beyond plastic pollution, Kiribati's response to climate change has been characterized by a range of innovative solutions. For example, the country's traditional wisdom, passed down through generations, has been fused with modern engineering practices to develop seawalls and other infrastructure to protect against rising sea levels and extreme weather events. Additionally, Kiribati has been working to preserve its cultural heritage by promoting traditional practices such as canoe building and fishing, which are essential to the country's way of life.

At the same time, Kiribati has been advocating for global collaboration to combat the impacts of climate change. The country has been vocal in international forums, calling attention to the unique challenges faced by small island developing states and pushing for more ambitious climate action from larger nations. Kiribati's leadership in this area highlights the importance of cross-border collaboration in tackling the global challenge of climate change.

In conclusion, Kiribati's response to climate change is a testament to the country's resilience and innovative spirit. Despite facing a range of challenges, the country has developed a range of community-driven initiatives, blended traditional wisdom with modern solutions, and advocated for global collaboration. The story of Kiribati is one of determination, hope, and the pursuit of a sustainable future, even in the face of an uncertain climatic landscape.

Chapter 3

Coping with Rising Tides and Extreme Events

Kiribati, a small island nation located in the Pacific Ocean, is facing a critical challenge due to the accelerated pace of coastal erosion and land loss caused by rising sea levels. The nation comprises 33 low-lying coral atolls and islands spread over 3.5 million square kilometers of ocean, making it one of the most vulnerable countries to the impacts of climate change.

The vulnerability of coastal communities in Kiribati is intensifying due to extreme weather events such as cyclones and storm surges, which increase the risk of displacement, infrastructure damage, and threats to food security and livelihoods.

The impending threat demands urgent measures to protect against coastal erosion and address the immediate challenges faced by vulnerable populations.

Kiribati is implementing adaptive measures to fortify coastal resilience. The nation is constructing seawalls, planting coastal vegetation, and exploring innovative engineering solutions to mitigate erosion. These measures are aimed at protecting homes, infrastructure, and agricultural land from the impacts of rising sea levels and extreme weather events.

Moreover, Kiribati is exploring relocation strategies and seeking international support and solidarity to build resilient communities and prepare for potential displacement. The nation is working with its development partners to secure funding for adaptation projects, capacity building, and infrastructure development.

The impacts of climate change pose a significant threat to the future of Kiribati. The nation faces an existential threat from rising sea levels, which could lead to the complete submergence of its islands and atolls. The challenges faced by Kiribati are not only environmental but also social and economic.

The displacement of coastal communities could lead to social and economic disruptions, affecting the well-being and livelihoods of thousands of people. Furthermore, the loss of land and agricultural productivity could impact food security and exacerbate poverty in the region.

To address these challenges, Kiribati is taking a comprehensive approach to building resilience. The nation is investing in education and awareness-raising programs to promote sustainable lifestyles and climate change adaptation. Kiribati is also developing its renewable energy sector to reduce greenhouse gas emissions and enhance energy security.

In addition, Kiribati is exploring innovative solutions to adapt to the impacts of climate change. The nation is researching the potential of floating islands, which could provide an alternative habitat for its population in case of displacement. Kiribati is also exploring the potential of seaweed farming as a climate-resilient source of livelihood for its coastal communities.

In conclusion, Kiribati's efforts to build resilience against the impacts of climate change are commendable. However, the nation requires urgent support and solidarity from the international community to address the challenges it faces. The world must come together to support vulnerable nations such as Kiribati in their efforts to adapt to the impacts of climate change and build a sustainable future for all.

Chapter 4

Biodiversity and Threats to Marine Life

Climate change is causing significant damage to the marine biodiversity of Kiribati, a small island nation in the Pacific Ocean. The warming of the oceans is leading to widespread coral bleaching, which poses a threat to the vibrant and diverse coral reef ecosystems. Coral reefs are home to a variety of marine life, including fish, crustaceans, and mollusks. They play a vital role in the ocean's ecosystem by providing habitats, shelter, and food for many marine creatures. The bleaching event disrupts the delicate balance of marine life, affecting the fish populations and endangering the traditional fishing practices that are crucial for Kiribati's food security.

The implications of this bleaching event extend beyond just the loss of biodiversity. It can affect the entire marine food web and the sustainability of Kiribati's coastal communities. Fish is the primary source of protein for the Kiribati people, and they rely heavily on traditional fishing practices. The loss of fish populations due to coral bleaching has a devastating impact on the food security of the country. It also affects the livelihoods of the fishing communities and the tourism industry that depends on the beauty of the coral reefs.

Kiribati has taken several measures to counter the impact of climate change on its marine biodiversity. The government has established marine protected areas, where fishing is restricted to allow the coral reefs to recover. They are also promoting sustainable fishing practices, such as using selective fishing gear that reduces bycatch and minimizes the damage to the coral reefs. However, there are still challenges in effectively enforcing regulations and curbing illegal fishing activities. The remote location of Kiribati and the vastness of its waters make it difficult to monitor and enforce regulations effectively.

Therefore, collaboration with regional and international partners is essential to safeguard Kiribati's marine biodiversity and ensure sustainable fisheries management for future generations. The government of Kiribati has been working closely with international organizations such as the United Nations Development Program (UNDP) and the Pacific Community (SPC) to develop strategies for marine conservation and sustainable fishing. The government has also signed several regional agreements on fisheries management and conservation, such as the Nauru Agreement and the Parties to the Nauru Agreement (PNA).

In conclusion, climate change is posing a significant threat to the marine biodiversity of Kiribati. Coral bleaching caused by warming oceans is disrupting the delicate balance of marine life and affecting the fish populations that are crucial for the food security of the country. Kiribati has taken several measures to counter the impact of climate change on its marine biodiversity, such as establishing marine protected areas and promoting sustainable fishing practices. However, collaboration with regional and international partners is essential to safeguard Kiribati's marine biodiversity and ensure sustainable fisheries management for future generations.

Chapter 5

Water Scarcity and Freshwater Security

Kiribati, a small island nation in the Pacific Ocean, is currently facing a severe freshwater scarcity crisis. The crisis is primarily due to saltwater intrusion and depletion of groundwater reserves. The contamination of freshwater sources by saltwater makes it unfit for consumption, thereby posing a serious health risk to the people of Kiribati.

The situation in Kiribati has been exacerbated by the erratic precipitation patterns, which have made the adoption of rainwater harvesting a challenging yet vital alternative. In recent years, the nation has experienced a decline in rainfall, leading to a decline in the availability of freshwater sources.

To address this issue, Kiribati has adopted innovative strategies such as desalination projects and community-driven water conservation initiatives. The desalination projects are aimed at converting seawater into freshwater for consumption. Community-driven water conservation initiatives are also being implemented to educate the locals on the importance of conserving water and adopting sustainable water management practices.

Rainwater harvesting techniques are being modernized and revitalized, with new technologies being developed to harvest rainwater more efficiently. Traditional water preservation methods are also being adapted to cope with the escalating water scarcity.

The focus has shifted towards sustainable water management practices to ensure that freshwater security is maintained. Kiribati's government is working with international organizations to develop strategies that will ensure the sustainable management of water resources.

The freshwater scarcity crisis in Kiribati has had a signifi-cant impact on the people of the nation. The lack of access to clean freshwater has led to an increase in waterborne diseases, creating a public health emergency. The government of Kiri-bati has recognized the need for urgent action to address the water crisis and is taking steps to mitigate the situation.

One of the government's primary goals is to ensure that the people of Kiribati have access to clean and safe drinking water. This has led to the development of innovative water manage-ment projects, such as the construction of water tanks and the installation of water filtration systems.

The government is also working with international organi-zations to develop strategies that will address the root cause of the water crisis. These strategies include the promotion of sustainable agriculture practices and the implementation of policies that will reduce the amount of water wastage.

The people of Kiribati are also playing an active role in addressing the water crisis. Community-led initiatives, such as the construction of rainwater harvesting systems, are being implemented to ensure that households have access to clean and safe water.

In conclusion, Kiribati's freshwater scarcity crisis is a com-plex issue that requires a multifaceted approach. The govern-ment of Kiribati is taking steps to address the crisis, and the people of the nation are actively participating in the efforts to mitigate the situation. The adoption of sustainable water management practices is crucial to ensure that the nation's freshwater security is maintained in the long term.

Chapter 6

Energy Transition and Sustainable Practices

Kiribati is a remote island nation that faces significant challenges with its energy supply. It is heavily reliant on imported fossil fuels, making it vulnerable to supply chain disruptions and high energy costs. This, in turn, hinders economic development and access to reliable electricity across the islands. However, transitioning to renewable energy sources like solar power and wind energy becomes crucial for Kiribati's energy independence.

To address these challenges, Kiribati has initiated projects that promote energy efficiency and sustainability. These projects aim to reduce reliance on fossil fuels, ensure consistent electricity supply, and mitigate environmental impacts, fostering Kiribati's self-sufficiency in energy. The country has set itself an ambitious target of generating 100% of its electricity from renewable energy sources by 2025.

One of the innovative approaches adopted by Kiribati is the use of solar power. The country has abundant sunshine throughout the year, making it an ideal candidate for solar power. The government has set up solar power systems on the roofs of public buildings, such as schools and hospitals, to provide reliable electricity to these facilities. Additionally, Kiribati has established a solar farm, which is capable of generating 500 kilowatts of electricity, enough to power 400 homes. The country has also installed solar-powered street lights, reducing the reliance on fossil fuels for lighting.

Another renewable energy source that Kiribati is exploring is wind energy. The country has identified two sites for wind turbines, one on the island of Tarawa and the other on the island of Kiritimati. These sites have been selected based on their wind potential, and the turbines are capable of generating up to 1.5 megawatts of electricity each. The government has also introduced a program that encourages households to install small-scale wind turbines, promoting the use of wind energy at a grassroots level.

Kiribati's efforts to transition to renewable energy sources are not only beneficial for the country's energy independence but also for the environment. The use of fossil fuels contributes to greenhouse gas emissions, which cause climate change. By reducing reliance on fossil fuels, Kiribati is mitigating the environmental impacts of its energy use, contributing to global efforts to combat climate change.

In conclusion, Kiribati's remote location and heavy reliance on imported fossil fuels present significant challenges to its energy supply. However, the country's initiatives to promote renewable energy sources like solar power and wind energy are commendable. These projects aim to reduce reliance on fossil fuels, ensure consistent electricity supply, and mitigate environmental impacts, fostering Kiribati's self-sufficiency in energy. Kiribati's efforts are not only beneficial for the country's energy independence but also for the environment, contributing to global efforts to combat climate change.

Chapter 7

Preserving Cultural Heritage Amidst Change

The threat of climate change has become a major concern for many countries around the world. Kiribati is one of the countries that has been heavily impacted by climate change. The rising sea levels and intense weather conditions have threatened the cultural heritage of this small island nation. The preservation of traditional knowledge, practices, and languages has become increasingly challenging as ancestral lands erode and communities face relocation.

Kiribati's cultural heritage is deeply connected to the land and ocean. The sacred sites and cultural rituals that have been passed down from generation to generation face the risk of being lost forever. This erosion of cultural heritage not only affects Kiribati's identity but also impacts social cohesion and mental well-being within communities. The loss of cultural heritage can create a sense of disconnection and loss of identity, which can lead to social and psychological problems.

Despite the challenges, Kiribati showcases resilience in preserving its cultural heritage. Community-driven efforts focus on revitalizing traditional practices, documenting oral histories, and passing down ancestral knowledge to younger generations. The adaptation of cultural traditions to new circumstances signifies the determination of Kiribati's people to protect their heritage amidst changing realities. These efforts have helped to maintain a sense of cultural identity and connection to the land and ocean.

One of the most significant challenges facing Kiribati's cultural heritage is the loss of ancestral lands. As sea levels rise, many of the low-lying islands that make up Kiribati are being eroded or completely submerged. This loss of land has a profound impact on the cultural heritage of the country, as many sacred sites and cultural practices are tied to specific locations on the islands. The relocation of communities to higher ground further complicates the preservation of cultural heritage, as it requires the adaptation of traditional practices to new environments.

Despite these challenges, Kiribati has made significant progress in preserving its cultural heritage. Community-led initiatives have helped to document oral histories and revitalize traditional practices, which has helped to maintain a sense of cultural identity and connection to the land and ocean. The preservation of cultural heritage is critical for the well-being of

Kiribati's people, as it helps to maintain a sense of continuity and connection to the past.

In conclusion, the threat of climate change poses a profound risk to Kiribati's cultural heritage. The erosion of ancestral lands and the relocation of communities pose significant challenges to the preservation of traditional knowledge, practices, and languages. Despite these challenges, Kiribati showcases resilience in preserving its cultural heritage through community-driven efforts to revitalize traditional practices, document oral histories, and pass down ancestral knowledge to younger generations. The preservation of cultural heritage is critical for the well-being of Kiribati's people and helps to maintain a sense of cultural identity and connection to the land and ocean.

Chapter 8

Advocating for Global Climate Action

Kiribati, a small island nation in the Pacific Ocean, has emerged as a vocal advocate for global climate action. The country has been representing the voices of vulnerable nations in international forums and has been urging the world to take urgent climate action. Kiribati's leaders have been engaging in international dialogues and calling for ambitious mitigation and adaptation measures, which have been resonating with people across the globe.

The reason behind Kiribati's strong stance on climate change is the disproportionate impact of climate change on low-lying island nations like Kiribati. Rising sea levels, severe weather events, and ocean acidification have been threatening Kiribati's very existence. The country's leaders have been emphasizing the urgent need for action to curb greenhouse gas emissions and to ensure that global warming does not exceed 1.5 degrees Celsius above pre-industrial levels. They have also been calling for financial assistance and support for vulnerable nations to adapt to and mitigate the effects of climate change.

Kiribati's efforts to raise awareness about the impact of climate change have been commendable. The country has been sharing its experiences and the challenges it faces with the world, highlighting the need for global collaboration and solidarity to address climate change. Kiribati has stressed the importance of fulfilling global commitments, providing financial assistance, and supporting vulnerable nations in their efforts to tackle climate change.

The stories of shared experiences and cooperation between nations facing similar challenges have underscored the importance of unity in the face of a common threat. Kiribati has become a symbol of hope for the world, showing that even small nations can make a big impact on the global stage. The country's efforts have inspired people around the world to take action on climate change and to work towards a more sustainable future.

In conclusion, Kiribati's emergence as a vocal advocate for global climate action is a testament to the country's resilience and determination. The nation has been amplifying its call for urgent climate action, representing the voices of vulnerable nations in international forums, and emphasizing the disproportionate impact of climate change on low-lying island nations like Kiribati. The world needs to take heed of Kiribati's call for action and work together to address the greatest threat facing humanity.

Chapter 9

Health Impacts and Community Well-being

Kiribati is a small island nation in the Pacific Ocean that is grappling with a range of health challenges that are being exacerbated by climate change. The country is made up of 33 coral atolls, which are home to around 110,000 people. Rising temperatures are causing heat-related illnesses, while water scarcity and contamination pose a threat of waterborne diseases. The mental health impacts of displacement, loss of cultural identity, and uncertainty about the future are also having a significant impact on community well-being.

The rising temperatures in Kiribati are causing a range of health problems. Heat-related illnesses are becoming more common, especially among vulnerable groups such as the elderly and young children. The increase in temperature is also leading to an increase in diseases such as dengue fever and malaria, which are spread by mosquitoes. The lack of clean water is also a problem in Kiribati, with many people relying on rainwater for drinking and cooking. The shortage of clean water is leading to an increase in waterborne diseases such as typhoid and cholera.

In addition to physical health problems, Kiribati is also facing significant mental health challenges. The loss of cultural identity, displacement, and uncertainty about the future are having a profound impact on the mental health of the community. Many people are experiencing anxiety, depression, and trauma as a result of these challenges. The lack of access to mental health services is also a problem in Kiribati, with many people struggling to access support.

Despite these challenges, Kiribati is taking steps to protect public health. The government is working to implement adaptive measures that will help to safeguard the health of the community. Community health education programs are being run that focus on practices that are resilient to climate change, preventing diseases, and providing mental health support. These programs are helping people to understand the risks they face and the steps they can take to protect themselves.

The country is also developing its health infrastructure to increase its resilience against health risks that are caused by climate change. This includes improving access to clean water and sanitation facilities, as well as investing in health care facilities and services. Kiribati is also collaborating with international health organizations to access resources and expertise that will help them to build resilience against climate-induced health risks.

In conclusion, Kiribati is facing a range of health challenges that are being worsened by climate change. Rising temperatures are causing heat-related illnesses, while water scarcity and contamination pose a threat of waterborne diseases. The mental health impacts of displacement, loss of cultural identity, and uncertainty about the future are also having a significant impact on community well-being. Despite these challenges, Kiribati is taking steps to protect public health through community health education programs, health infrastructure development, and collaboration with international health organizations.

Chapter 10

Education, Innovation, and Future Prospects

Climate change is a global phenomenon that affects every aspect of life, including education. Kiribati, a small island nation in the Pacific, is one of the countries most vulnerable to the impacts of climate change. The nation's location makes it particularly susceptible to sea-level rise, ocean acidification, and extreme weather events such as cyclones. These changes have a significant impact on the country's education system.

Displacement and environmental changes are two major issues that affect schooling in Kiribati. The rise in sea levels and the increased frequency of cyclones have caused many families to relocate from their homes. As a result, children are forced to change schools, disrupt their education, and lose access to quality education. The environmental changes, such as flooding and erosion, also damage school infrastructure, making it difficult for children to attend classes and access educational resources.

However, Kiribati recognizes the importance of education and its role in building a resilient and sustainable future. The government has implemented innovative educational initiatives to improve access to quality education. One such initiative is the integration of climate change-focused curriculum in schools. This curriculum aims to educate students about the impacts of climate change and methods to mitigate its effects. The curriculum equips students with the knowledge and skills to adapt to the challenges posed by climate change.

Another initiative is the use of technology-based learning. The use of digital technologies, such as online learning platforms and e-books, allows students to access educational resources even in remote areas. This approach not only addresses the issue of access but also prepares students for the digital age and the future of work.

Kiribati's government also focuses on nurturing future generations amidst climate uncertainties. The nation invests in innovative technologies, vocational training, and sustainable livelihood programs to empower youth to become resilient agents of change. The use of innovative technologies, such as solar energy, promotes sustainable development and addresses the energy challenges faced by the nation. Vocational training equips students with the skills needed for the job market, enabling them to secure employment and contribute to the country's economy. Sustainable livelihood programs promote sustainable agriculture and fishing practices, ensuring the nation's food security and reducing its dependency on imports.

In conclusion, climate change poses a significant threat to education in Kiribati. However, the government's innovative initiatives, including climate change-focused curriculum integration and technology-based learning, pave the way for adaptive education systems. Kiribati's commitment to fostering innovation and sustainability underscores its determination to build a more resilient and prosperous future for everyone.

Chapter 11

Governance and Climate Policy

The funding allocations that Kiribati receives enable the nation to leverage international support for climate-resilient projects and sustainable development. This is an important aspect of Kiribati's efforts to address the impacts of climate change and ensure a sustainable future for its people.

It is important to note that Kiribati advocates for environmental justice, which involves addressing issues of fairness and equity in climate change impacts. The nation emphasizes the disproportionate burden borne by vulnerable populations and highlights the need for inclusive policies and global cooperation to ensure environmental justice.

This approach is aligned with the global efforts to address climate change and its impacts. Environmental justice is an important aspect of climate action, as it recognizes the need to address the social and economic inequalities that often exacerbate the impacts of climate change on vulnerable populations.

In Kiribati's case, the nation's small size and limited resources make it particularly vulnerable to the impacts of climate change. Therefore, it is important for the nation to advocate for policies and programs that prioritize environmental justice and ensure that the benefits of climate action are shared equitably among its people.

Kiribati also advocates for indigenous rights in climate change discourse, urging recognition of indigenous knowledge and rights in international climate agreements. This is an important aspect of the nation's efforts to address the impacts of climate change on its people and ensure a sustainable future for its indigenous communities.

Indigenous knowledge and rights are critical to addressing the impacts of climate change, as they provide a unique perspective on the relationship between human societies and the natural environment. By recognizing and respecting indigenous knowledge and rights, we can develop more effective strategies for addressing the impacts of climate change and ensuring a sustainable future for all.

In Kiribati, indigenous communities have a deep understanding of the natural environment and the ways in which it is changing due to climate change. Therefore, it is important for the nation to recognize and respect this knowledge and work with indigenous communities to develop effective strategies for addressing the impacts of climate change.

Overall, Kiribati's advocacy for environmental justice and indigenous rights in climate change discourse is an important aspect of the nation's efforts to address the impacts of climate change and ensure a sustainable future for its people. By prioritizing these issues, we can work towards a more just and equitable future for all.

Chapter 12

Economic Challenges and Adaptation Strategies

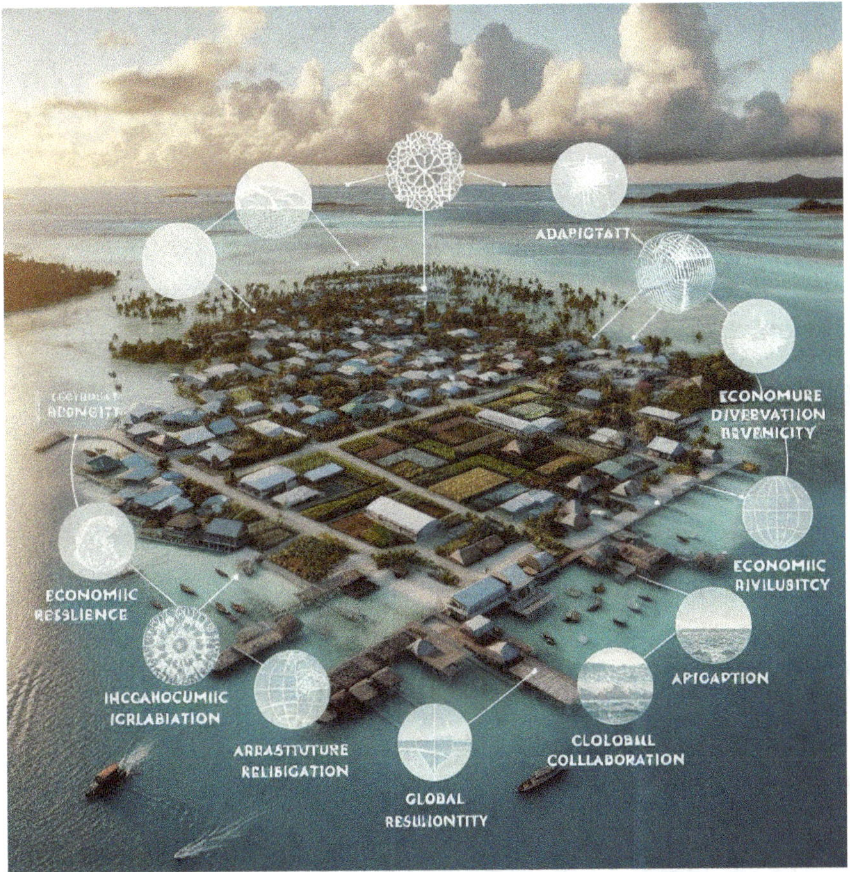

Kiribati, a Pacific island nation, confronts profound economic challenges exacerbated by the impacts of climate change. The nation's traditional livelihoods, primarily centered around fishing and agriculture, face disruption due to rising sea levels and extreme weather events. These climate-induced changes directly threaten the foundation of Kiribati's economy, creating vulnerabilities that extend to the very essence of its societal structure.

The destruction of critical infrastructure, including ports and roads, compounds economic risks. Rising sea levels and intensified weather events contribute to coastal erosion, jeopardizing transportation networks and hindering the movement of goods and services. This infrastructure vulnerability further amplifies the economic instability faced by Kiribati, making recovery and adaptation more challenging.

Kiribati's economic landscape is characterized by limited diversification, with historical reliance on specific sectors. The narrow economic base leaves the nation particularly exposed to shocks within these key industries, as disruptions reverberate through the entire economy. The lack of alternative income sources exacerbates the challenges posed by climate change, contributing to the nation's economic fragility.

In addition to internal vulnerabilities, Kiribati's economy is susceptible to external shocks. The nation's small and open economy leaves it at the mercy of global economic trends. Fluctuations in international markets, particularly in vital sectors like fisheries, directly impact Kiribati's economic stability. The country's dependence on foreign aid introduces an additional layer of vulnerability, as changes in aid levels or conditions can influence fiscal stability and the ability to address economic challenges effectively.

The social and economic implications of these vulnerabilities are profound. Income inequality within Kiribati's population may widen as those reliant on the most affected sectors experience heightened hardships. Job losses resulting from disruptions in traditional livelihoods contribute to social and economic stress, affecting the overall well-being of individuals and families.

Addressing Kiribati's economic vulnerability requires a comprehensive and integrated approach. Sustainable economic diversification, resilience-building measures, and international collaboration are essential components of a strategy aimed at mitigating the impacts of climate change and fostering long-term economic sustainability. Balancing immediate economic needs with the imperative of long-term adaptation remains a critical challenge for Kiribati's policymakers in navigating the complexities of their economic vulnerability.

In response to the profound economic challenges posed by climate change, Kiribati has adopted a strategic focus on economic resilience and adaptation. Central to this approach is a commitment to diversifying the nation's economy, steering away from traditional sectors vulnerable to climate-related disruptions. This deliberate shift reflects an acknowledgment of the need for a more robust and adaptable economic foundation.

A key component of Kiribati's economic resilience strategy involves the promotion of sustainable tourism. Recognizing the potential of this sector to generate revenue and employment opportunities, the nation aims to leverage its natural beauty and unique cultural heritage to attract visitors. By fostering sustainable tourism practices, Kiribati seeks to build a resilient economic sector that can withstand the impacts of climate change while contributing to the nation's overall economic stability.

In addition to tourism, Kiribati is actively working on fostering economic sectors less susceptible to the adverse effects of climate change. By identifying and supporting industries with inherent resilience, the nation aims to create a diversified economic portfolio that can absorb shocks and maintain stability in the face of environmental challenges. This forward-looking approach involves a careful assessment of the strengths and vulnerabilities of various sectors, guiding strategic investments and policies.

The initiatives to enhance food security play a crucial role in Kiribati's economic resilience and adaptation efforts. Innovative agricultural practices are being explored to address the challenges posed by climate-induced changes, such as shifts in precipitation patterns and saltwater intrusion. By embracing sustainable and adaptive farming techniques, Kiribati aims to secure a stable food supply, mitigating the risks associated with disruptions to traditional agricultural practices.

Furthermore, Kiribati is actively exploring alternative livelihoods that align with the nation's economic and environmental goals. By diversifying the range of opportunities available to its population, Kiribati aims to provide individuals and communities with more options to thrive amidst changing economic and environmental landscapes. This approach not only enhances the nation's economic resilience but also fosters adaptability at the individual and community levels.

In conclusion, Kiribati's focus on economic resilience and adaptation reflects a proactive response to the challenges posed by climate change. Through strategic economic diversification, promotion of sustainable tourism, support for resilient sectors, and initiatives to enhance food security and explore alternative livelihoods, Kiribati is laying the groundwork for a more resilient and adaptive economic future. These efforts underscore the nation's commitment to navigating the complexities of climate-induced economic vulnerabilities and building a sustainable foundation for its people.

Chapter 13

Gender Dynamics and Climate Resilience

The gendered impacts of climate change in Kiribati reveal a complex and nuanced set of challenges, with women experiencing disproportionate effects due to their roles in households and communities. As climate change alters environmental conditions, women find themselves at the forefront of coping with these changes, particularly in the context of household management. The traditional responsibilities of women in Kiribati, including securing food, water, and maintaining the well-being of the family, become increasingly challenging as climate-related disruptions impact these crucial aspects of daily life.

One of the prominent challenges faced by women is related to water collection. Changes in precipitation patterns and the increased frequency of extreme weather events can affect water availability, necessitating longer and more arduous journeys to collect water. This places an additional burden on women, who are often responsible for ensuring an adequate and safe water supply for their families. The gendered nature of water collection exacerbates existing gender inequalities and underscores the unique challenges faced by women in adapting to changing environmental conditions.

Moreover, the increased care burdens during displacement further contribute to the gendered impacts of climate change. As communities grapple with the consequences of sea-level rise and extreme weather events, displacement becomes a reality for many. In such situations, women often shoulder a disproportionate share of caregiving responsibilities, caring for children, elderly family members, and the sick. The strain on women's time and energy can limit their capacity to engage in income-generating activities or community resilience efforts, reinforcing gender disparities in the face of climate-induced challenges.

However, it is essential to recognize the pivotal roles that women play in community resilience and adaptation efforts. Despite facing distinct challenges, women in Kiribati are active agents of change within their communities. Their knowledge of local ecosystems, traditional coping mechanisms, and community networks positions them as key contributors to adaptive strategies. Women's involvement in decision-making processes related to climate change adaptation is crucial for developing effective and inclusive responses that consider the unique needs and perspectives of both genders.

In conclusion, the gendered impacts of climate change in Kiribati illuminate the differential vulnerabilities experienced by women. While facing distinct challenges related to household management, water collection, and increased care burdens during displacement, women also emerge as essential actors in community resilience and adaptation efforts. Recognizing and addressing the gender-specific dimensions of climate change impacts is vital for fostering inclusive and effective strategies that enhance the resilience of Kiribati's communities in the face of environmental challenges.

In Kiribati, a concerted effort is underway to integrate gender-responsive strategies into climate adaptation and policy formulation, recognizing the critical role that women play in building climate resilience. The emphasis on empowering women encompasses multifaceted approaches, ranging from education and leadership opportunities to targeted interventions designed to address gender-specific vulnerabilities.

Education stands out as a foundational element of empowering women in the context of climate resilience. By providing women with access to education and training programs, Kiribati aims to equip them with the knowledge and skills needed to actively engage in climate adaptation initiatives. Education becomes a tool for enhancing women's understanding of environmental changes, sustainable practices, and their own agency in contributing to community resilience.

Leadership roles for women are actively promoted within the framework of gender-responsive strategies. Recognizing the importance of diverse perspectives in decision-making processes related to climate adaptation, Kiribati seeks to increase the representation of women in leadership positions. Empowering women to take on leadership roles not only ensures their voices are heard but also enriches the adaptive strategies with a more comprehensive understanding of community needs.

Targeted interventions tailored to address gender-specific vulnerabilities are a key component of Kiribati's gender-responsive approach. These interventions recognize and address the unique challenges faced by women, such as increased care burdens, limited access to resources, and specific health concerns. By tailoring adaptation measures to the specific needs of women, Kiribati aims to enhance the effectiveness and inclusivity of its climate response strategies.

Inclusive policies and community programs are pivotal in advancing gender equality in climate response. Kiribati acknowledges the importance of considering gender-specific perspectives and experiences in the formulation of policies and the implementation of community-based programs. Ensuring that both women and men actively participate in and benefit from climate adaptation initiatives contributes to a more equitable and resilient society.

By adopting gender-responsive strategies, Kiribati not only acknowledges the distinct challenges faced by women but actively works to create an environment where women are empowered to contribute significantly to climate resilience. These strategies align with broader goals of inclusivity, equality, and sustainable development, reflecting a commitment to building a resilient and gender-responsive framework for addressing the impacts of climate change in Kiribati.

Chapter 14

Community Resilience and Cultural Innovations

In Kiribati, community resilience is deeply rooted in the rich tapestry of traditional knowledge, cultural values, and robust community networks. The resilience exhibited by Kiribati's communities serves as a testament to the strength derived from the interplay of these elements, forming the bedrock of their ability to navigate environmental uncertainties.

Traditional knowledge plays a pivotal role in community resilience, as it draws upon the wisdom passed down through generations. Kiribati's communities possess a deep understanding of their local ecosystems, weather patterns, and sustainable practices that are intricately linked to their cultural heritage.

This knowledge becomes a valuable resource in adapting to changing environmental conditions, enabling communities to make informed decisions based on centuries-old insights.

Cultural values form another cornerstone of community resilience in Kiribati. These values foster a strong sense of identity, unity, and collective responsibility within communities. In the face of environmental uncertainties, the adherence to cultural values becomes a source of strength, guiding community members in their actions and interactions. This shared cultural foundation contributes to a cohesive and resilient community fabric.

Community networks, characterized by strong interpersonal relationships and mutual support, play a crucial role in building resilience. These networks create a system of shared responsibility, where community members actively collaborate to address challenges and share resources. In times of environmental uncertainties, the interconnectedness within these networks allows for the efficient dissemination of information, collective decision-making, and the pooling of resources to respond to emerging needs.

The fostering of community cohesion is a deliberate effort to strengthen the social fabric that underpins resilience. Kiribati recognizes the importance of creating spaces for community members to come together, share experiences, and collectively strategize for the future. Community cohesion not only enhances the emotional well-being of individuals but also forms a foundation for coordinated action in the face of environmental challenges.

Shared knowledge becomes a catalyst for collective action, wherein communities collaboratively develop and implement strategies to adapt and thrive in the midst of uncertainties. This collective approach empowers communities to leverage their combined resources, skills, and traditional wisdom to address both immediate and long-term challenges, fostering resilience that goes beyond individual capabilities.

In conclusion, Kiribati's approach to community resilience building embraces the symbiotic relationship between traditional knowledge, cultural values, and community networks. By fostering community cohesion, shared knowledge, and collective action, Kiribati's communities stand resilient in the face of environmental uncertainties, embodying the strength derived from their rich cultural heritage and interconnected social fabric.

In Kiribati, cultural innovations for adaptation represent a harmonious integration of traditional wisdom and modern technologies, showcasing the resilience of the community in the face of climate change. Leveraging indigenous knowledge, Kiribati actively incorporates time-tested practices with contemporary approaches, resulting in innovative adaptation strategies that cater to local needs and emphasize the value of cultural heritage in climate resilience.

One notable cultural innovation is the adoption of agroforestry practices. Drawing upon traditional agricultural knowledge, Kiribati communities combine the cultivation of diverse crops with strategically planted trees to create sustainable and resilient agroecosystems. Agroforestry not only enhances food security but also contributes to soil conservation, water management, and biodiversity preservation. This cultural innovation reflects a dynamic adaptation strategy that integrates ancient farming practices with contemporary agricultural techniques.

Traditional water conservation methods play a crucial role in Kiribati's adaptation efforts. Indigenous knowledge related to water harvesting and storage is combined with modern technologies to optimize water resources in the face of changing climatic conditions. Kiribati communities design and implement systems that capture and store rainwater, ensuring a sustainable and reliable water supply. This cultural innovation not only addresses the challenges posed by water scarcity but also aligns with the community's historical practices of responsible resource management.

Architectural designs rooted in indigenous knowledge contribute to climate resilience by adapting to local environmental conditions. Traditional building techniques that account for natural ventilation, local materials, and elevation to mitigate flooding are integrated with modern construction practices. These culturally informed architectural innovations enhance the resilience of communities against the impacts of climate change, providing safer and more sustainable living spaces.

The integration of traditional wisdom with modern technologies in Kiribati's adaptation strategies reflects a dynamic and holistic approach to climate resilience. By recognizing the value of indigenous practices, the community emphasizes the importance of cultural heritage in shaping effective solutions to environmental challenges. This innovative blend of the old and the new not only showcases the adaptability of Kiribati's culture but also serves as a model for sustainable development that respects and builds upon traditional knowledge.

Chapter 15

Infrastructure Challenges and Resilience

Kiribati's critical infrastructure stands at the forefront of vulnerability, grappling with imminent risks amplified by climate change-induced hazards. The island nation faces escalating challenges as key components of its infrastructure, such as roads, water supply systems, and coastal facilities, become increasingly susceptible to the adverse impacts of climate change, including erosion, flooding, and extreme weather events.

Roads, essential arteries for transportation and connectivity, are under the looming threat of climate change-induced hazards. Rising sea levels and increased frequency of extreme weather events contribute to coastal erosion and flooding, directly impacting the integrity of road networks. As these vital transportation lifelines deteriorate, communities face disruptions in mobility, emergency response, and access to essential services.

Water supply systems, crucial for sustaining life and promoting public health, are under heightened vulnerability. The encroachment of saltwater due to rising sea levels poses a direct threat to freshwater resources, jeopardizing the availability of safe drinking water. Increased variability in precipitation patterns further exacerbates challenges in maintaining a reliable and resilient water supply, amplifying the vulnerability of Kiribati's communities to water scarcity and contamination.

Coastal facilities, including ports and infrastructure along the shoreline, confront escalating risks. Rising sea levels and intensified storm surges expose these critical assets to erosion and damage. The implications extend beyond the immediate impact on maritime activities, affecting trade, transportation, and the overall economic resilience of the nation.

The vulnerability of Kiribati's critical infrastructure underscores the urgent need for comprehensive adaptation measures to enhance resilience. Strategic planning, incorporating climate-resilient design and engineering practices, is imperative to safeguarding vital infrastructure components. This includes elevating structures, reinforcing coastal defenses, and implementing sustainable water management practices to mitigate the impacts of climate change.

Furthermore, community engagement and capacity-building efforts are essential to ensure the effective implementation of adaptation strategies. By involving local communities in resilience-building initiatives and providing them with the tools and knowledge to address climate risks, Kiribati can enhance the adaptive capacity of its population and reduce the overall vulnerability of critical infrastructure.

In conclusion, the vulnerability of Kiribati's critical infrastructure to climate change-induced hazards necessitates a proactive and multifaceted approach. Addressing these challenges requires a combination of resilient infrastructure development, sustainable water management, and community-based adaptation strategies to build a more robust and climate-resilient foundation for the nation's future.

In response to the escalating vulnerability of its critical infrastructure, Kiribati is actively pursuing resilient infrastructure solutions that encompass innovative engineering approaches and nature-based strategies. Recognizing the imperative of adapting to the impacts of climate change, the nation is undertaking infrastructure development initiatives designed to enhance resilience and minimize climate-related risks.

One key facet of Kiribati's approach involves the implementation of innovative engineering solutions tailored to withstand the challenges posed by climate change-induced hazards. This includes the development of climate-resilient coastal protection measures to safeguard against erosion and storm surges. By adopting advanced engineering practices, Kiribati aims to fortify its coastlines, protecting vital infrastructure assets from the increasing threats associated with rising sea levels and extreme weather events.

Nature-based approaches play a pivotal role in Kiribati's resilient infrastructure solutions. The nation recognizes the inherent value of ecosystems in providing natural defenses against climate impacts. Sustainable construction practices that integrate nature-based features, such as mangroves and coral reefs, are being explored to enhance coastal resilience. These approaches not only contribute to the protection of infrastructure but also foster biodiversity, supporting ecosystem health.

Infrastructure upgrades are a central component of Kiribati's resilience-building strategy. By incorporating climate-resilient features into existing and new infrastructure projects, the nation seeks to future-proof essential assets against the intensifying impacts of climate change. This includes the integration of sustainable materials, improved drainage systems, and elevated structures to mitigate the risks of flooding and erosion.

In addition to physical infrastructure enhancements, Kiribati is investing in capacity-building initiatives and community engagement to ensure the effective implementation of resilient solutions. Local communities are actively involved in the planning and execution of infrastructure projects, fostering a sense of ownership and empowerment. This participatory approach contributes to the overall success of resilience-building efforts, as communities become integral partners in adapting to climate challenges.

By focusing on climate-resilient infrastructure development, Kiribati aims to create a foundation that can withstand the evolving impacts of climate change. The combination of innovative engineering solutions, nature-based approaches, and community engagement reflects a comprehensive strategy to build a more robust and adaptive infrastructure network. In doing so, Kiribati endeavors to secure the well-being of its population, protect critical assets, and ensure sustainable development in the face of a changing climate.

Chapter 16

International Support and Climate Financing

Kiribati encounters formidable challenges in securing the necessary climate financing to support its crucial adaptation and mitigation efforts in the face of climate change. The nation grapples with a combination of factors that collectively impede its ability to access adequate financial resources from international climate funds.

A primary challenge lies in the limited financial resources available to Kiribati. The nation, characterized by its relatively small economy and population, faces constraints in generating the required funds to address the multifaceted impacts of climate change. The scarcity of domestic resources poses a significant barrier to independently financing comprehensive adaptation and mitigation projects.

The complexity of application procedures for international climate funds presents another hurdle for Kiribati. The intricate requirements and extensive documentation demanded by these funds create administrative challenges for the nation. Navigating through intricate application processes can be resource-intensive, requiring technical expertise and institutional capacity that Kiribati may find difficult to mobilize.

Competing priorities further compound the challenge of securing climate financing for Kiribati. The nation must allocate its limited resources across various sectors, including healthcare, education, and infrastructure, in addition to climate-related initiatives. The need to balance diverse priorities can divert attention and resources away from climate resilience projects, hindering the nation's ability to implement comprehensive climate adaptation and mitigation strategies.

Additionally, the vulnerability of Kiribati's critical infrastructure to climate change-induced hazards places a strain on the nation's financial resources. Urgent infrastructure upgrades and resilience-building measures require substantial investments, and the availability of climate financing becomes pivotal in addressing these pressing needs.

Despite these challenges, Kiribati remains committed to overcoming barriers to climate financing. The nation actively engages in international climate negotiations to advocate for simplified and more accessible funding mechanisms. Additionally, efforts to enhance local capacity, streamline application processes, and prioritize climate-related initiatives within national development plans are underway to strengthen Kiribati's ability to access and effectively utilize climate finance.

In conclusion, the challenges faced by Kiribati in securing climate financing are rooted in limited financial resources, complex application procedures, and competing priorities. Addressing these challenges requires concerted efforts at both national and international levels, emphasizing the need for streamlined processes, increased financial support, and a holistic approach to climate resilience that aligns with the nation's sustainable development goals.

In its pursuit of climate financing, Kiribati strategically engages in partnerships with international entities and multilateral organizations to secure the necessary support for climate-resilient projects and sustainable development. These collaborative efforts, encompassing various approaches such as strategic partnerships, capacity building initiatives, and advocacy for increased funding allocations, underscore Kiribati's commitment to mobilizing resources and expertise from the global community.

Strategic partnerships play a pivotal role in Kiribati's approach to climate financing. By fostering alliances with international partners, the nation gains access to a broader spectrum of resources, including financial support, technical expertise, and knowledge exchange. Collaborative initiatives enable Kiribati to leverage the strengths and capabilities of its partners, ensuring a more comprehensive and effective response to the challenges posed by climate change.

Capacity building initiatives constitute another key component of Kiribati's strategy. By enhancing its institutional and technical capacities, the nation is better equipped to navigate complex application processes and effectively manage climate finance resources. Through training programs, knowledge-sharing, and skill development, Kiribati builds a foundation for sustainable project implementation and long-term climate resilience.

Advocacy for increased funding allocations is integral to Kiribati's efforts in securing climate financing. The nation actively participates in international climate negotiations and forums, advocating for a fair and proportionate allocation of funds to address the specific challenges it faces. By voicing its concerns and highlighting the urgent need for support, Kiribati seeks to garner international recognition and commitment to financing initiatives that align with its climate resilience and sustainable development goals.

Furthermore, Kiribati aligns its climate financing strategies with its national development plans, emphasizing the integration of climate resilience into broader development objectives. This ensures that climate-related projects contribute to the nation's overall sustainable development agenda, fostering a synergistic approach that maximizes the impact of international funding.

In conclusion, Kiribati's partnerships and funding strategies for climate resilience are characterized by a multifaceted and collaborative approach. By engaging with international partners, building local capacity, and advocating for increased funding, Kiribati positions itself to effectively address the challenges of climate change. Through these concerted efforts, the nation strives to create a resilient and sustainable future that benefits both its citizens and the global community.

Chapter 17

Environmental Justice and Indigenous Rights

Kiribati is a staunch advocate for environmental justice, actively addressing issues of fairness and equity in the context of climate change impacts. The nation emphasizes the imperative of recognizing and rectifying the disproportionate burden borne by vulnerable populations, underlining the need for inclusive policies and global cooperation to ensure environmental justice.

Central to Kiribati's advocacy is the recognition that the impacts of climate change are not evenly distributed, with vulnerable and marginalized communities facing a disproportionately higher burden. These populations, often characterized by limited resources and resilience, bear the brunt of climate-related hazards, exacerbating existing social and economic disparities. Kiribati asserts the importance of acknowledging these disparities and actively working towards equitable solutions that prioritize the needs and rights of the most vulnerable.

In advocating for environmental justice, Kiribati places a strong emphasis on the development and implementation of inclusive policies. The nation contends that policies addressing climate change impacts must consider the unique vulnerabilities and challenges faced by different segments of the population. By prioritizing inclusivity, Kiribati seeks to ensure that the benefits of climate resilience measures are accessible to all, particularly those who are most adversely affected.

Global cooperation stands at the forefront of Kiribati's approach to environmental justice. Recognizing that climate change is a shared global challenge, the nation underscores the importance of collaborative efforts to mitigate its impacts. Kiribati actively engages in international forums, advocating for solidarity and collective responsibility in addressing climate change. The nation calls for developed countries to take leadership roles in supporting vulnerable nations and communities, fostering a sense of shared responsibility for environmental justice on a global scale.

Furthermore, Kiribati's advocacy extends to the promotion of climate justice within international climate negotiations. The nation actively participates in discussions aimed at shaping global climate policies that are fair, transparent, and responsive to the needs of those most affected by climate change. By championing the cause of environmental justice,

Kiribati contributes to a more equitable and sustainable approach to climate action.

In conclusion, Kiribati's commitment to environmental justice is evident in its advocacy for fairness and equity in addressing climate change impacts. By highlighting the disproportionate burden on vulnerable populations, emphasizing inclusive policies, and championing global cooperation, Kiribati seeks to contribute to a more just and resilient future for all, regardless of their socio-economic status or geographical location.

Kiribati takes a principled stance in advocating for indigenous rights within the discourse of climate change. The nation underscores the importance of recognizing and respecting the rights of indigenous communities, emphasizing the need for the incorporation of indigenous knowledge in international climate agreements. Kiribati actively engages in global forums, playing a vocal role in voicing the rights and concerns of indigenous communities facing climate-related challenges.

A key aspect of Kiribati's advocacy is the call for the acknowledgment of indigenous knowledge in climate change discussions. The nation contends that indigenous communities possess valuable insights, practices, and traditions that have enabled them to adapt to environmental changes over generations. Kiribati emphasizes the significance of integrating this indigenous wisdom into climate policies and strategies, fostering a more comprehensive and effective approach to addressing the impacts of climate change.

In international climate agreements, Kiribati actively promotes the inclusion of provisions that safeguard the rights of indigenous communities. The nation advocates for measures that ensure the participation of indigenous representatives in decision-making processes related to climate action. This includes the recognition of indigenous perspectives on adaptation and mitigation strategies, contributing to more culturally sensitive and locally relevant climate policies.

Kiribati's active participation in global forums is integral to amplifying the voices of indigenous communities facing climate-related challenges. By voicing the concerns and rights of these communities, Kiribati seeks to garner international support and solidarity for the protection of indigenous rights in the face of climate change. The nation leverages its presence in these forums to raise awareness about the vulnerabilities and unique circumstances of indigenous populations, fostering a greater understanding of their role in climate resilience.

Moreover, Kiribati emphasizes the need for climate policies that uphold the rights of indigenous communities to access and manage their traditional lands and resources. The nation advocates for measures that prevent the displacement and dispossession of indigenous populations due to climate-induced impacts, safeguarding their cultural heritage and connection to the land.

In conclusion, Kiribati's advocacy for indigenous rights in climate change discourse reflects a commitment to inclusivity and cultural sensitivity. By actively participating in global forums, emphasizing the incorporation of indigenous knowledge, and advocating for the rights of indigenous communities, Kiribati contributes to shaping a more equitable and respectful approach to climate action on the international stage.

Chapter 18

Technological Innovations for Climate Resilience

   Kiribati faces notable challenges in the realm of technology, encompassing infrastructure limitations, disparities in access to advanced technology, and shortages in a skilled workforce. These challenges collectively impact crucial aspects such as data collection, analysis, and the establishment of early warning systems for natural disasters.

   One prominent challenge is the limited technology infrastructure in Kiribati. Insufficient technological infrastructure hinders the nation's capacity to deploy and sustain advanced technological solutions, particularly in communication networks that affect the seamless transfer of data and information, especially in times of emergency.

Another challenge is the disparities in access to advanced technology. The availability of cutting-edge technological tools and equipment is constrained, limiting the nation's ability to leverage innovative solutions for climate adaptation, disaster risk reduction, and sustainable development. This lack of access contributes to a technological gap affecting various sectors of Kiribati's economy.

The shortage of a skilled workforce poses a considerable challenge as well. The expertise required to effectively manage and operate advanced technologies, particularly in the context of climate monitoring and disaster response, is limited. This shortage hampers Kiribati's capacity to harness the full potential of available technologies and impedes the development of a workforce capable of addressing technological challenges.

These technological constraints directly impact data collection and analysis mechanisms. Kiribati faces difficulties in implementing comprehensive data collection systems for climate-related parameters, hindering the accurate assessment of environmental changes. Inadequate technology infrastructure further complicates the analysis of collected data, limiting the depth of insights into climate trends and impacts.

The absence of robust technology infrastructure poses challenges in establishing effective early warning systems for natural disasters. Timely and accurate dissemination of information is crucial for community preparedness and response. Technological limitations impede the development and maintenance of early warning systems that can withstand the environmental challenges faced by Kiribati.

Due to the gaps in technology, Kiribati often relies on traditional methods for various activities. While traditional knowledge and practices are valuable, they may not provide the level of precision and efficiency offered by modern technologies. This reliance on traditional methods could pose limitations in responding to rapidly changing climate conditions and potential hazards.

Despite these challenges, opportunities for improvement exist. Collaborative efforts with international partners, investments in technology infrastructure, and targeted capacity-building programs can enhance Kiribati's technological capabilities. Leveraging innovative solutions tailored to Kiribati's context, such as low-cost and sustainable technologies, presents opportunities for addressing specific challenges and fostering resilience.

In conclusion, Kiribati's technology challenges encompass infrastructure limitations, disparities in access to advanced technology, and skilled workforce shortages, impacting critical areas such as data collection, analysis, and the establishment of early warning systems. Addressing these challenges requires a multi-faceted approach, including strategic investments, international collaboration, and the exploration of innovative, context-specific solutions to enhance Kiribati's technological resilience.

Significant progress has been made in technological solutions within Kiribati, marking a key step in the nation's adaptation efforts. The integration of traditional knowledge with modern technology stands out as a noteworthy approach. This integration is exemplified by specific advancements, such as the utilization of satellite imagery to monitor coastline changes. By combining indigenous wisdom with technological tools, Kiribati enhances its capacity to track and respond to environmental shifts, particularly those influenced by climate change.

The implementation of climate-resistant infrastructure emerges as another vital aspect of technological adaptation. Kiribati recognizes the importance of constructing infrastructure that can withstand the impacts of climate change, such as rising sea levels and extreme weather events. This proactive approach to building resilient structures aligns with the nation's commitment to mitigating the risks posed by a changing climate and ensuring the long-term sustainability of its infrastructure.

Moreover, ongoing projects in Kiribati emphasize community engagement as a crucial element of technology adoption. The nation understands that for technological solutions to be effective, they must be embraced and utilized at the local level. These projects prioritize the empowerment of local communities, ensuring that they not only participate in but also benefit from technological innovations. This community-centric approach fosters a sense of ownership and responsibility, contributing to the sustainable integration of technology into Kiribati's adaptation strategies.

In conclusion, Kiribati's journey in technological solutions and adaptation reflects a harmonious blend of traditional knowledge and modern technology. From satellite imagery for coastline monitoring to climate-resistant infrastructure, these advancements showcase the nation's commitment to leveraging innovation for climate resilience. The emphasis on community engagement further underscores Kiribati's strategic approach, ensuring that technological solutions are not only adopted but embraced at the grassroots level, thereby enhanc-ing the nation's overall adaptive capacity.

Chapter 19

Community-Led Climate Action and Empowerment

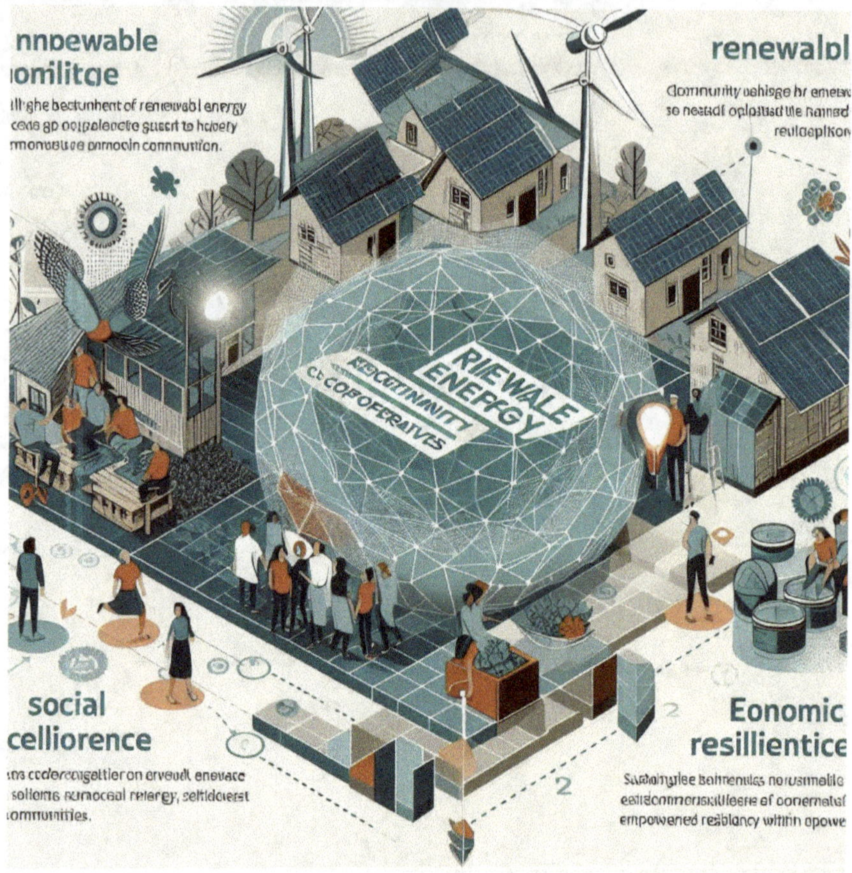

Community engagement plays a pivotal role in Kiribati's climate action efforts, as highlighted in various grassroots initiatives that empower local communities. One key aspect is the organization of local workshops centered on sustainable practices. These workshops serve as platforms for community members to learn about and implement environmentally friendly approaches. By disseminating knowledge at the grassroots level, Kiribati ensures that communities are equipped with the tools and information needed to actively contribute to climate action.

Another crucial component of community engagement involves community-based disaster risk reduction programs. Kiribati recognizes the vulnerability of its communities to climate-related hazards and actively involves them in designing and implementing measures to reduce these risks. This collaborative approach not only enhances the effectiveness of risk reduction strategies but also instills a sense of resilience within the communities, fostering the capacity to withstand and recover from climate-induced disasters.

Participatory decision-making processes in climate adaptation planning further underscore Kiribati's commitment to involving local communities. By including community members in the decision-making process, the nation ensures that adaptation strategies align with the specific needs and priorities of each community. This inclusive approach not only enhances the relevance and effectiveness of climate adaptation plans but also fosters a sense of ownership among community members.

Success stories abound in Kiribati, showcasing communities actively contributing to climate action through their engagement in various initiatives. These stories highlight the transformative impact of empowering communities to take charge of their environmental destiny. As individuals and groups within communities actively participate in climate action, a collective sense of ownership and responsibility emerges, creating a foundation for sustained efforts in building resilience and addressing the challenges posed by climate change.

In conclusion, community engagement is a cornerstone of Kiribati's approach to climate action. From sustainable practice workshops to community-based disaster risk reduction programs and participatory decision-making, the nation recognizes the indispensable role of local communities in shaping and implementing effective climate adaptation strategies. Through these initiatives, Kiribati not only strengthens its resilience to climate change but also cultivates a shared commitment to environmental stewardship at the grassroots level.

Community empowerment in Kiribati extends beyond awareness to tangible sustainable practices, exemplifying a holistic approach to addressing climate challenges. Resilient agricultural methods stand out as a prominent example of how empowered communities actively contribute to sustainability. The implementation of drought-resistant crop cultivation techniques showcases the practical application of knowledge acquired through community empowerment initiatives. By adopting agricultural practices that withstand climatic stresses, communities enhance their food security and contribute to the overall resilience of the agricultural sector.

Waste management campaigns represent another impactful sustainable practice driven by empowered communities. Through the voluntary efforts of community members, these campaigns address environmental concerns by promoting responsible waste disposal and recycling. The active involvement of volunteers not only contributes to cleaner and healthier surroundings but also fosters a sense of collective responsibility for environmental stewardship. Such initiatives demonstrate the power of community empowerment in driving positive environmental outcomes.

The establishment of renewable energy cooperatives further illustrates Kiribati's commitment to sustainable practices initiated by empowered communities. By coming together to create cooperatives, community members harness the potential of renewable energy sources. This not only contributes to reducing the carbon footprint but also strengthens local economies by promoting sustainable energy alternatives. The cooperative model emphasizes collaboration, ensuring that the benefits of renewable energy initiatives are shared among community members, enhancing economic resilience in the face of climate challenges.

These examples underscore the interconnectedness of sustainable practices, social cohesion, and economic resilience within empowered communities. By actively engaging in initiatives like resilient agriculture, waste management campaigns, and renewable energy cooperatives, communities in Kiribati not only address climate challenges but also build a foundation for sustainable development. The holistic approach to community empowerment contributes not only to environmental sustainability but also to the overall well-being and resilience of Kiribati's communities.

Chapter 20

Future Aspirations and Global Collaboration

Kiribati envisions a sustainable future characterized by a holistic approach that addresses climate challenges while fostering resilience and preserving cultural heritage. A central component of this vision is the development of climate-resilient infrastructure. The page emphasizes the importance of infrastructure that can withstand the impacts of climate change, including rising sea levels and extreme weather events. By investing in resilient infrastructure, Kiribati aims to ensure the long-term viability of essential facilities and services, promoting stability and security for its communities.

Educational reforms take center stage in Kiribati's vision for a sustainable future, with a specific focus on climate literacy. Recognizing the critical role of education in building awareness and empowering communities, the nation strives to integrate climate-related knowledge into its educational curricula. By fostering a deep understanding of climate issues among its citizens, Kiribati aims to cultivate a proactive and informed populace capable of contributing to sustainable practices and resilience-building efforts.

Preserving cultural heritage emerges as another key goal in Kiribati's vision for the future. Innovative projects are envisioned to safeguard and celebrate the rich cultural legacy of the nation. These projects aim to strike a balance between embracing modernity and preserving traditional practices, ensuring that Kiribati's unique identity thrives amidst the challenges of a changing climate. The integration of cultural preservation into the sustainability vision reflects the nation's commitment to maintaining a strong connection to its roots.

The page further explores how Kiribati translates these aspirations into tangible actions through the integration of goals into national policies and strategies. By embedding sustainability objectives into governance frameworks, Kiribati seeks to institutionalize its commitment to a sustainable future. This strategic approach ensures that the envisioned goals become integral components of the nation's development trajectory, guiding decisions and initiatives toward a resilient, climate-literate, and culturally rich future.

Kiribati is actively engaged in global collaboration and solidarity efforts to address the formidable challenge of climate change. The nation participates in various international forums, diplomatic initiatives, and partnerships, demonstrating its commitment to collective action. Kiribati plays a proactive role in advocating for global climate justice, emphasizing the necessity of fair and equitable responses to the challenges posed by climate change.

A significant focus is on diplomatic endeavors to rally support for vulnerable nations. Kiribati effectively communicates the unique challenges it faces due to climate change, leveraging diplomatic channels to garner understanding and support from the international community. By fostering empathy and solidarity, Kiribati aims to build a coalition that collectively addresses the shared responsibility of mitigating and adapting to climate change.

The nation is actively involved in fostering cross-border cooperation, recognizing that climate change is a global challenge requiring collaborative solutions. Kiribati engages with other countries, sharing knowledge, expertise, and best practices. Through these partnerships, Kiribati contributes to a collective pool of wisdom that enhances the effectiveness of climate action strategies. This collaborative approach underscores the understanding that a united front is crucial for addressing the multifaceted and interconnected issues posed by climate change.

In summary, Kiribati's dedication to global collaboration and solidarity is a vital component of its strategy to combat climate change. Through active participation in international forums, diplomatic initiatives, and partnerships, Kiribati seeks to amplify its voice, advocate for climate justice, and contribute to a collective effort aimed at securing a sustainable and resilient future for all nations.

Chapter 21

Environmental Education and Awareness

Environmental education holds paramount importance in Kiribati, playing a crucial role in fostering awareness, inducing behavioral changes, and promoting sustainable practices. This page sheds light on the pivotal role of environmental education as a catalyst for positive change within the nation.

One key aspect highlighted is the integration of environmental education into school curricula. Recognizing the formative role of education, Kiribati prioritizes the inclusion of environmental topics in schools. This ensures that the younger generation is equipped with the knowledge and understanding necessary to address environmental challenges. By instilling environmental consciousness at an early age, Kiribati aims to nurture a generation that values and actively contributes to environmental sustainability.

Additionally, the page delves into the implementation of community workshops as a means of disseminating environmental knowledge. These workshops serve as platforms for engaging with local communities, sharing information on sustainable practices, and fostering a sense of collective responsibility towards the environment. The participatory nature of these workshops encourages active involvement, empowering individuals to take ownership of environmental stewardship.

Moreover, the importance of public awareness campaigns and outreach initiatives is underscored. By reaching out to the broader population through various channels, Kiribati aims to raise awareness about environmental issues and promote responsible behavior. This multifaceted approach acknowledges the diverse ways in which individuals access information and ensures a comprehensive strategy for disseminating environmental knowledge.

In conclusion, the page emphasizes the multifaceted role of environmental education in Kiribati. From school curricula to community workshops and public outreach, the nation recognizes the transformative power of education in shaping attitudes and behaviors towards the environment. This concerted effort aligns with Kiribati's commitment to building a sustainable and environmentally conscious society.

This section delves into the theme of empowering individuals through knowledge, with a specific focus on success stories arising from environmental education initiatives in Kiribati. The narrative unfolds by highlighting the transformative effects of educational programs, showcasing how individuals gain not only knowledge but also a heightened awareness of the environment and its intricacies. This newfound awareness becomes a catalyst for a sense of responsibility and stewardship, instilling in individuals a commitment to contribute positively to climate resilience and conservation efforts.

The narrative further illustrates success stories that underscore the practical outcomes of environmental education. It may delve into specific examples of community-led conservation projects, sustainable practices adopted by individuals, or innovative solutions devised through the application of acquired knowledge. These success stories serve as inspiring examples, showcasing the tangible and positive changes brought about by empowering individuals through environmental education.

Additionally, the narrative may explore how these empowered individuals become advocates for environmental causes within their communities. The ripple effect of knowledge-sharing and empowerment is emphasized, portraying a community that actively engages in climate resilience and conservation efforts. This grassroots involvement becomes a testament to the transformative power of knowledge in creating positive change.

In conclusion, the section highlights the success stories emerging from environmental education initiatives in Kiribati, portraying empowered individuals as champions of environmental stewardship, contributing to the broader goals of climate resilience and conservation within the nation.

Chapter 22

Adaptive Governance and Policy Innovation

Exploring the hurdles associated with adaptive governance, this content delves into the complexities faced in implementing policies and governance structures tailored for climate adaptation in Kiribati. It sheds light on critical challenges such as institutional capacity, underlining the necessity of robust institutions capable of effectively responding to climate-related issues. The narrative further navigates the intricacies of policy coherence, illustrating the difficulties in aligning diverse policies to create a cohesive and efficient climate governance framework.

Additionally, the content explores stakeholder engagement as a pivotal element of adaptive governance, emphasizing the challenges related to ensuring active participation and collaboration among various stakeholders. It underscores the need for inclusive decision-making processes and may discuss the difficulties in balancing the interests of different sectors and communities. Fostering partnerships is highlighted as essential for successful climate governance.

The content may incorporate specific examples or case studies to illustrate governance challenges faced by Kiribati, providing instances where institutional limitations, policy incoherence, or insufficient stakeholder engagement hindered the effectiveness of climate governance initiatives. These real-world examples add depth to the understanding of the challenges and underscore the urgency of addressing them for improved adaptive governance.

In summary, the content outlines the multifaceted challenges in adaptive governance faced by Kiribati, emphasizing critical areas like institutional capacity, policy coherence, and stakeholder engagement. The overarching message underscores the importance of addressing these challenges to enhance the nation's ability to adapt to and mitigate the impacts of climate change.

Detailing the innovative strides in climate policy within Kiribati, this content showcases the nation's forward-thinking approaches to address the challenges of climate change. The narrative highlights key policy measures that reflect Kiribati's commitment to resilience and adaptation.

One notable aspect discussed is the integration of traditional knowledge into legislative frameworks. The content explores how Kiribati has recognized the value of indigenous wisdom in tackling climate-related issues. This integration not only reflects cultural sensitivity but also enhances the effectiveness of policies by incorporating time-tested practices.

The content further delves into community-driven climate action plans as a novel approach adopted by Kiribati. It illustrates how local communities actively participate in shaping and implementing policies tailored to their specific needs and vulnerabilities. This decentralized approach fosters a sense of ownership and ensures that policies resonate with the diverse contexts within the nation.

Inclusive decision-making processes are another key facet emphasized in this content. It outlines how Kiribati prioritizes the involvement of various stakeholders, including communities, in shaping climate policies. This inclusivity aims to ensure that diverse perspectives contribute to robust policy formulations and implementations.

The overarching theme throughout the content is Kiribati's dedication to policy innovation in the face of climate change. By incorporating traditional knowledge, engaging communities, and fostering inclusivity, the nation aims to create adaptive and effective policies that resonate with its unique context. This narrative paints a picture of Kiribati as a proactive player in the global effort to address climate challenges through innovative policy measures.

Chapter 23

Resettlement and Future Scenarios

This content delves into the intricate issue of resettlement, shedding light on the challenges, ethical considerations, and social impacts entwined with relocation resulting from climate-induced displacement. The narrative navigates through various perspectives, encompassing community concerns and the imperative preservation of cultural identity throughout the resettlement processes.

One primary focus is on the challenges inherent in the resettlement endeavors. The content outlines the multifaceted obstacles faced by communities undergoing relocation, ranging from the disruption of established livelihoods to the strain on existing social structures. It highlights the complexities involved in orchestrating resettlement initiatives and the need for comprehensive strategies to address diverse challenges.

Ethical considerations form a crucial component of the content, exploring the moral implications associated with climate-induced displacement and resettlement. It delves into the ethical dimensions of decision-making, emphasizing the importance of fair and just practices in navigating the sensitive terrain of relocating communities. The narrative underscores the ethical responsibility of stakeholders in ensuring the well-being and rights of affected populations.

The content also delves into the social impacts of resettlement, examining how the process can shape and transform communities. It considers the potential effects on social cohesion, cultural heritage, and community dynamics, emphasizing the significance of safeguarding these elements amidst the challenging backdrop of relocation.

Throughout the narrative, various perspectives are woven together to provide a comprehensive understanding of the complexities involved in resettlement. From community-level concerns to overarching ethical considerations, the content aims to foster awareness and nuanced discussions surrounding climate-induced displacement and the associated challenges in navigating resettlement processes.

Continuing the discussion, the content further delves into the intricate aspects surrounding the issue of resettlement in Kiribati. It elaborates on the multifaceted challenges, ethical considerations, and social impacts that come into play when communities face relocation due to climate-induced displacement. The aim is to provide a comprehensive understanding of the various perspectives involved, including the concerns expressed by affected communities.

Moreover, the content explores Kiribati's proactive approach in implementing innovative and inclusive policy measures to tackle the complexities associated with resettlement. It emphasizes the significance of legislative frameworks that integrate traditional knowledge, community-driven climate action plans, and decision-making processes that actively engage stakeholders. By spotlighting these policy innovations, the content underscores Kiribati's dedication to adapting to climate change while safeguarding the rights and well-being of its communities.

In summary, the content strives to offer an in-depth exploration of the challenges and considerations linked to resettlement, presenting a balanced perspective on the ethical, social, and policy dimensions involved in addressing climate-induced displacement in Kiribati.

Chapter 24

Reflections and Moving Forward

This reflective segment serves as a poignant summary of the profound narrative embedded within the pages of the book, shedding light on Kiribati's resilient journey in the face of relentless climate change impacts. Through the lens of reflection, the narrative unfolds, capturing the essence of the nation's unwavering spirit in confronting and overcoming multifaceted challenges.

The pages of this book vividly depict Kiribati's resilience as the nation grapples with the far-reaching consequences of climate change. From the intricate challenges of economic vulnerabilities to the intricate web of gendered impacts, the narrative weaves together a tapestry of collective strength and determination. Each chapter unfolds as a testament to the nation's ability to not only endure but to stand firm and united against the adversities that climate change presents.

At its core, the reflective exploration underscores the interconnectedness of Kiribati's communities, drawing upon traditional knowledge, cultural values, and collaborative networks as foundational pillars of resilience. The narrative unfolds with a focus on the indomitable spirit of the people, whose shared experiences and collective action form the bedrock of their ability to withstand environmental uncertainties.

The reflective pages encapsulate the broader message that Kiribati offers to the world — a message of hope, unity, and the power of collective action in the face of climate change. It invites readers to ponder not only the challenges faced by Kiribati but also the invaluable lessons in resilience, adaptation, and the enduring strength found within the bonds of community and shared purpose.

In drawing the final curtain on this literary journey, the concluding chapter serves as a poignant reflection on Kiribati's formidable odyssey in confronting the relentless impacts of climate change. Through the narrative, the resilience, determination, and collaborative spirit of the Kiribati people emerge as a triumphant force against adversity. As the pages turn, readers are invited to immerse themselves in the emotional and inspiring tapestry woven by Kiribati's experiences.

This closing chapter is not merely an endpoint but a transition, providing a bridge to the future. It recognizes that the battle against climate change is an ongoing narrative, demanding continuous efforts and unwavering commitment. The narrative extends beyond the confines of the book, encouraging readers to become active participants in the ongoing struggle to safeguard the planet and its vulnerable communities.

Envisioning the future becomes a central theme, with the chapter emphasizing the paramount importance of sustained collaboration. The challenges posed by climate change are dynamic and multifaceted, requiring adaptive and unified approaches. It articulates the necessity for continuous innovation, underscoring the role of creative solutions and adaptive strategies in navigating the complex and evolving landscape of climate change.

At the heart of this concluding chapter lies the call for global solidarity. The narrative acknowledges that the impacts of climate change transcend geographical boundaries, necessitating a collective and interconnected response. It advocates for a world where nations, communities, and individuals stand together to address the shared challenges of a changing climate. The chapter resonates with a call to action, urging readers to actively engage in the collective endeavor of safeguarding the planet for both present and future generations.

This closing segment serves as an optimistic epilogue, leaving readers with a profound sense of purpose and agency. It instills a forward-thinking mindset, encouraging individuals to be proactive contributors to a sustainable and resilient future. As the book gracefully concludes, the narrative emphasizes the transformative power of collaboration, innovation, and global solidarity as the driving forces charting the path ahead for Kiribati and, by extension, for the entire world.